BEEKEEPING EQUIPMENT

(caveat emptor)

*A monograph for the
initiated and the uninitiated*

BEEKEEPING EQUIPMENT
(caveat emptor)

*A monograph for the
initiated and the uninitiated*

Prepared by:
J.D.YATES B.Sc. (Hons), C.Eng., FIEE.

NORTHERN BEE BOOKS

First published 2002
Second Edition 2009

COPYRIGHT © J.D.Yates
ISBN 978-1-904846-40-6

Printed and bound in Great Britain by
Lightning Source UK

Design and Artwork
D&P Design and Print
Set in Adobe Jenson Pro

Published by Northern Bee Books
Scout Bottom Farm,
Mytholmroyd,
Hebden Bridge
West Yorkshire HX7 5JS

CONTENTS

Preface   Page viii

Foreword to the 2nd edition   Page x

Introduction   Page 1
     The Migration of the honeybee   Page 1
     Characteristics of *Apis mellifera*   Page 3
     Propolis   Page 4
     The BBKA & non-existent standards   Page 5
     Units of measurement   Page 6

Bee space   Page 7
     Historical   Page 7
     What we now understand by bee space   Page 8
     Brace comb, burr comb and propolis   Page 9
     Design criteria for hives and hive parts   Page 10

Equipment that fails the design criteria   Page 12
     Hives   Page 12
     Hive - frame runners   Page 12
     Queen excluders   Page 15
     Metal/plastic ends and other spacers   Page 17
     Frames general   Page 19
     Brood frames   Page 20
     Super frames   Page 27

Other equipment requiring a re-think   Page 31
     Smoker and hive tools   Page 31
     Hive tools   Page 31
     Copper smoker   Page 32
     Floorboards   Page 37

Entrance blocks/mouse guards ..... Page 37
Roofs ..... Page 39
Crown (Cover) boards ..... Page 41
Clearer boards ..... Page 41
Feeders ..... Page 42
Butler cage ..... Page 42
Extractors ..... Page 43
Brood boxes ..... Page 43
Beeswax foundation ..... Page 44
Porter be escapes ..... Page 46
Other items ..... Page 47

Proposals ..... Page 48
Standards and the organisation ..... Page 48
Committees and those on them ..... Page 51

Conclusions ..... Page 52

Imperial/metric conversions ..... Page 53

List of drawings

Migration of the honeybee	Page 2
Grecian hive with removable top bars	Page 7
Runners - long lug hives	Page 13
Runners - short lug hives	Page 13
Queen excluders -wire type	Page 16
Wild comb - cross section	Page 21
Hoffman spacing	Page 22
Top bars	Page 24
Standard brood frame	Page 25
Manley type spacing	Page 29
Hive tool	Page 32
Smoker	Page 33
Entrance block/mouse guard	Page 38
BBKA membership	Page 50

PREFACE

After starting beekeeping during the second World War, I have seen no real advance or improvement in the design and manufacturing quality of beekeeping equipment. If anything, a decline has taken place over the last 60 years or so which can be partially explained as a result of cost cutting. I am, of course, talking generally and there are occasional exceptions to the general trend. There are one or two manufactures that are likely to dispute hotly my views but this matters little if, in the future, an improvement can be achieved.

The trouble arises because those designing and manufacturing the equipment are either inexperienced beekeepers or they don't keep bees at all on a regular basis. A further factor aggravating the situation is the discontinuation of standards and specifications for the manufacturer to work to and the disbandment of those organisations with a responsibility for making and publishing them. In this country the onus rests squarely on the BBKA.

I have given many lectures on the inadequacy of the equipment available for sale and very few demurring words have been said about the content of these talks. In fact, it is the many requests for written notes of these talks that has prompted the writing of this monograph. Significant improvements can be made to existing equipment with very little effort and expense providing a short term solution, while the long term solution can also be achieved without undue expenditure.

I have written the few words that follow for the many existing beekeepers that could make their beekeeping life a whole lot easier by adopting some of the short term modifications. It is written for those starting beekeeping as a hobby and for all the other beekeepers that care for the future and would wish to see a noticeable improvement in the equipment available. The diagrams have been inserted as close to the relevant text as possible for easy reading. The originals of these diagrams were initially in colour but in order to keep the production costs to a minimum they have been reproduced in black and white; long experience of beekeepers has made me aware of their short arms and deep pockets. I shall argue later that the only way that improvement can be effected, in the long term, is a determined effort by the BBKA.

I trust that the future will see an easement in the task of beekeeping husbandry by the provision of better equipment more suited to the job it is meant to do as, surprisingly, little has changed over the last 100 years.

J.D.Yates, Newton Ferrers, Devon. 2002

FOREWORD to the 2nd EDITION

I have looked over the few words that I wrote in 2002 and find that nothing of any moment has changed except that there has been an increase in the membership of the BBKA. Additionally, mesh floorboards have become popular together with monitoring boards, both of which can be regarded as still under development.

The same old equipment is still being manufactured with the same inherent defects and the BBKA have done nothing to address the problem of standards. In fact, there has been no comment from them about my proposals. The review of the 1st edition by David Cramp, the Editor of Apis-UK, rather predicted such a negative approach by the BBKA.

I can only hope that the few equipment manufacturers remaining in the UK can get together with the national beekeeping body and make a little progress in this matter which has laid stagnant for 6 years since I first brought the matter to their attention. It would cost peanuts for the equipment manufacturers to tool up and produce the parts for, say, a standard frame complying with all the design criteria.

J.D.Yates, Newton Ferrers, Devon. 2009

INTRODUCTION

There are two important aspects of beekeeping husbandry which prompt the requirement for well designed equipment, sadly, both have been overlooked in most of the classical literature. They are the adverse temper of most bees in the United Kingdom, and many other places in Europe, coupled with the fact that many of these bees produce propolis in greater or lesser quantities depending on their genetic make-up. Good beekeeping husbandry requires the manipulation of bee colonies throughout the active season and bad tempered bees coupled with propolis make these every-day manipulations difficult to perform. Before addressing any possible solutions it is essential that we have a full understanding of the problems.

It is important to understand how bad temper arises and to comprehend this the origin of the honeybee and its migration throughout the world must be examined together with an understanding of why the temperaments of our honeybee is so variable. There is a very wide range of temperaments from the very docile to the very aggressive, or put another way, from those that exhibit little defensive instinct to those with a very strong defensive instinct whether the colony is disturbed or remains undisturbed.

Migration of the honeybee.

By various scientific techniques, including palaeontology, it is now generally agreed that the origin of the honeybee was in that part of Africa that is now known as Kenya, somewhere between 20 and 30 million years ago. This was before India split off from Africa and before the Red Sea and Rift Valley were formed. The land mass, now called India, carried with it some of the earlier bee-like insects which developed into the Eastern species of bees, ie. *Dorsata, Cerana* and *Florea*.

Reference to the diagram of 'The Migration of the Honeybee', shows the origin in Kenya and the main migration routes marked with double line arrows northwards, southwards and westwards. Major races developed notably *Apis Mellifera capensis* in the Cape of Good Hope area, *Adansonii* in the equatorial strip, *Fasciata* (Egyptian bee) in the north east and *Intermissa* (Tellian or Arab bee) in the north west.

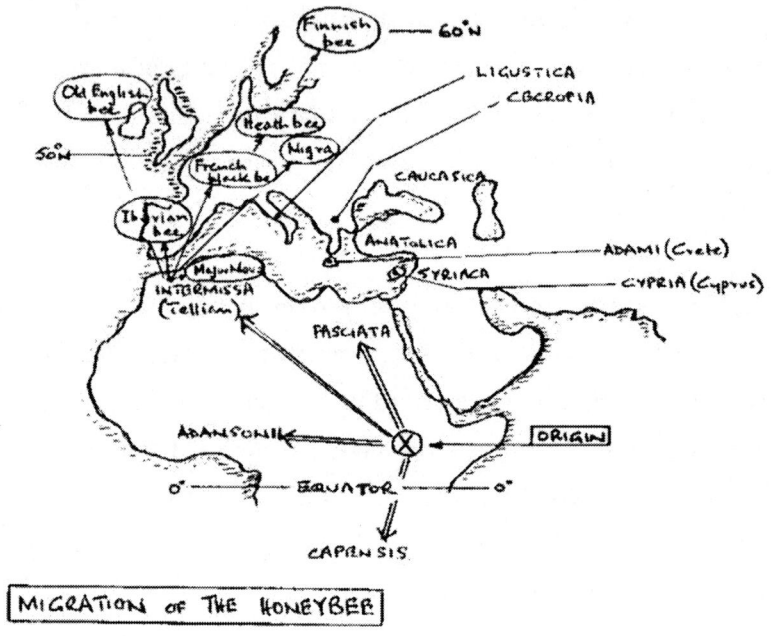

The Tellian bee is considered to be a major race from which many other strains have developed. In its native part of northwest Africa it developed in a very hostile environment and, presumably, adapted to these conditions by defending its nest against determined predators (eg. hornets, etc.). It migrated northwards before the last melting of the ice cap c. 10,000 years ago and established the well known races in NW Europe such as the Iberian bee, the French Black bee and the English bee. It migrated as far north as Finland in latitude 60°N.

Similarly, the other major race (*Fasciata*) migrated northwards providing the Italian bees (*Ligustica*), Greek, Caucasian, Carniolan, etc. After the melting of the ice cap, the UK was cut off from the continent and the Mediterranean was formed isolating Europe from Africa. At that time we had a discrete race of bee in the UK now known as the English Black bee or the English Brown bee. It was decimated by disease in the early 1900s (nb. Isle of Wight Disease or Acarine) and the Government of the day offered subsidies to beekeepers to import bees from the Continent. The result was that most of the races in Europe were imported and over the last 50 to 80 years these races have interbred and we now have a hotchpotch of mongrels with a very mixed pool of genes.

All the bees with their origins emanating from the Tellian bee have to a greater or lesser degree the defensive instinct of their ancestors while other bees from Italy, Greece, etc. have a very weak defensive trait. Breeding can only be true with pure strains, while breeding with mongrels is known to be an erratic and unpredictable procedure.

Herein then lies the root of our problem in the UK. Using mongrel bees and taking pot luck with the matings will result in a wide range of temperaments in the offspring. The only beekeeper who claimed experience with the old English bee who and has committed his findings to paper, was Bro. Adam. He considered the bee had some very desirable features but also some very bad ones, notably bad temper. BIBBA claim that there are pockets of the old English bee with a good temper and campaign for its breeding and use in the UK. To date, there is little tangible evidence of success after more than 25 years and the problem of bad temper prevails.

Characteristics of *Apis mellifera*.

Before leaving this introduction to understanding the temperament of our bees it must be pointed out that other characteristics will also be present and variable with mongrels such as high tendency to swarm, heavy propolis gathers, early starters/late finishers, longevity, economy on stores for overwintering, resistance to different diseases, etc. We know from experience that it is possible to eliminate bad temper fairly quickly by culling those queens producing bad tempered bees. However, it is nigh impossible to control more than two variables without recourse to specialised isolation and mating techniques. Our own breeding programme concentrates on good temper and minimum swarming tendency which can be achieved by any beekeeper if they put their mind to it. Unfortunately, it is a very small minority who control the temperament of the bees they keep and we put this down to two traits that have developed over the last 50 years as follows:

a) Newcomers have been encouraged to wear very adequate protective clothing which has become available making them 'safe' from stings under most conditions.

b) Few associations encourage their membership to manipulate their bees without gloves.

Propolis.

All honeybees collect propolis to a greater or lesser degree primarily for varnishing and re-varnishing the cells in the brood nest before their use by the queen and for filling in cracks in the nest or hive structure. Propolis has anti-bacterial qualities, hence its use in the brood nest. Without it, the brood nest would be a very unhealthy nursery with thousands of larvae defaecating in their cells prior to pupation. In warm weather it is sticky and in cooler weather conditions it becomes hard and brittle. Under both conditions it makes the manipulation of hive parts difficult without a hive tool which, unless used very carefully, jars the hive parts as they are separated.

The problem is now reasonable well defined; couple bad tempered bees with propolis and not only does any manipulation become difficult, it becomes most unpleasant. Many newcomers to the craft are put off at this stage and never restart, while others have to don impenetrable armour to tend their bees.

The solution to the bad tempered bees has been outlined above. A large part of what follows in later chapters will be addressing the propolis problem.

The BBKA and non-existent standards.

The BBKA were the responsible body for setting up the necessary standards and specifications for bees and beekeeping equipment in this country. They promulgated standards for bees and colonies and co-opted the services of the British Standards Institute (BSI) to evolve the standards for frames and equipment. In September/October 1984 the following British Standards were withdrawn:

1947 - Bees (colonies and nuclei)
1950 - Honey grading glasses
1960 - Bee hives, frames and wax foundation

With the result that now there are no nationally recognised standards for equipment used by beekeepers in the United Kingdom. This unfortunate state of affairs arose, so I was told by the BSI, because of lack of interest on the part of the BBKA who previously provided representatives on a special BSI committee dealing with the subjects quoted above. The BBKA side of the story is slightly different; the standards were abandoned on the grounds of cost. Both explanations amount to the same thing. The BBKA were probably right to limit the expenditure incurred by the BSI in 1984. However, they failed to provide and improve the standards and specifications themselves, thereby leaving the commercial manufacturers to produce some very unsatisfactory equipment. In the following chapters numerous examples of this will be examined.

The story so far looks pretty gloomy but it will cheer up as the saga unfolds!

Units of measurement.

M ost equipment has evolved and has been based on the Imperial System of measurements and, while not wishing to offer myself as a metric martyr, I shall use these units as the primary measure and include a separate table of conversions rather than showing the metric equivalent in parenthesis throughout the text. Please see page 53 where the metric equivalents have been calculated to one tenth of a milli-metre. For practical purposes these calculated equivalents can be further rounded up or down, for example 5/16 inch = 0.312 inch = 7.9 mm. This is the well known bee space and can be considered to be 8 mm.

It is to be noted that twist drills and other tools marketed in metric sizes may be obtained in sizes that increase in 0.1 mm steps, eg. drills of 2.6 mm, 2.7 mm diameter.

BEE SPACE

Historical.

Prior to the mid 1800s when bee space was first recognised all beekeeping was undertaken in skeps or similar types of hives where, we are told, it was not possible to remove the combs for inspection and management of the colony

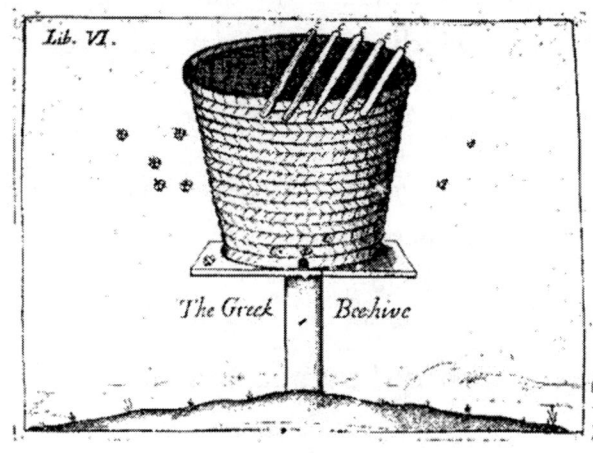

The first recorded hive whereby it was possible to remove the comb was in Grecian hives and described in a book by George Wheeler entitled "A journey into Greece" and published in 1682. The hive illustrated above (a copy of plate 127 from Herrod-Hempsall's book, Beekeeping New and Old volume 1) had bars across a tapered container. The bees built their comb on the underside of the bars which were removed by cutting the wax joining the combs to the tapered sides. Such a hive could not be regarded as a truly removable frame hive by today's standards but never-the-less it was a very advanced design at the time.

There is evidence that Major Munn designed a moveable frame hive in 1834 and received a French patent for it in 1838. It is not clear whether Munn's hive incorporated bee space. A further moveable frame hive designed by J.A. is described in a letter by him dated 16th June 1863; even to this day the name of J.A. remains unknown. The plans show that it had no bee space between the walls and the frames.

The Rev. L.L. Langstroth of Philadelphia USA takes credit for 'inventing' bee space although he did not personally claim this, he only recognised the need for it in the design of his hive which was patented on 5th October 1852 and the hive itself appeared in 1853.

After 1853 the bee keeping world really 'took off' and many inventions followed such as the queen excluder, smoker, wax foundation, extractors, etc. Modern beekeeping had begun.

What we now understand by bee space.

Rev.L.L.Langstroth used a space of ½ inch the design of his original hive as the clearance between the walls of the hive and the side bars of the frames. Today we know that bees will eventually, and under certain conditions, build comb in a space of this size.

By observation over many years by many expert beekeepers, it is now known that honeybees of the *Apis mellifera* species will build brace comb and burr comb in a space greater than ⅜ inch and will fill a space with propolis if it is less than ¼ inch.

Between these two extremes the honeybee will maintain this passage-way and not obstruct it with wax or propolis. The mid-dimension ⁵⁄₁₆ inch is now regarded as the correct dimension for bee space. This allows ± ¹⁄₁₆ inch for manufacturing tolerances and or the expansion or contraction of construction materials due to climatic conditions.

All the above measurements have been quoted in imperial measurements because all the popular hives have been designed using imperial measurements. Hives used in the United Kingdom include the National, Commercial, WBC and its variants, Smith, etc. and the most widely used hive in the world, the Langstroth hive, used in the USA, Australia and New Zealand. I have grave doubts about converting the measurements to metric but for the record here they are:

¼ inch = 6.35 mm, 5/16 inch = 7.9375 mm, ⅜ inch = 9.525 mm.

It is normal practice to round-up or round-down these metric conversions but should these be to the nearest 1.0 mm or to the nearest 0.5 mm? For equipment manufacturing purposes, using wood as the material, rounding to the nearest millimeter would be adequate giving 6 mm, 8 mm and 10 mm for the three dimensions shown above. Thus we can now say that:

Bee space = 5/16 inch ± ¹⁄₁₆ inch or 8 mm ± 2 mm.

Brace comb, burr comb and propolis.

There seems to be confusion about the terms brace comb and burr comb. 'Brace' means to hold something together generally in the horizontal plane whereas 'burr' is something which is proud, for example, the rough edges around a hole where the drill emerges. Applying this to comb, then brace comb is built horizontally between top bars when the spacing is incorrect and burr comb is built vertically, for example, joining the top bars of one box of frames to the bottom bars of the next box above when the spacing is incorrect. Some writers, for reasons unknown, equate brace and burr suggesting that they are one and the same thing. However, one thing is certain, comb will be built in spaces greater than ⅜ inch.

With spaces of less than ¼ inch the honeybee will fill it with propolis. There are exceptions to this which usually occur at the entrance to the hive where sometimes curtains are built using enormous amounts of propolis.

Design criteria for hives and hive parts.

Using the well known dimensions for bee space, it is now possible to define the criteria for design purposes which are as follows:

a) All spaces in the hive and its components shall be ⁵⁄₁₆ inch.
It is important to know where these spaces are situated. In a modern hive they are as follows:

i)   Between the two walls of the hive and the frame side bars.
ii) Between the other two walls of the hive and the face of the end combs.
iii) Between the top bars in one box and the bottom bars of the box above.
iv) Between the top bars and the under surface of the crown board
     or the Miller/Ashforth feeder.
v)   Between the sides of adjacent top bars.
vi) Between the sides of adjacent side bars.
vii) Between the underside of the frame lugs and the upper surface of the
     rebate or rabbet formed by the metal runners.

There is only one exception and that is the space below the frames in the lower brood box which is adjacent to the floor. This becomes ⁵⁄₁₆ inch (using bottom bee space construction) plus the depth of the floor (usually ⅞ inch) making a total of ¹³⁄₁₆ inches which accomodates queen cells or 'parked' bees in bad weather.

b) All spaces that are less than ¼ inch shall be minimised to prevent propolisation.
Again it is important to familiarise ourselves with the location of the points where trouble may occur. In a modern hive they are as follows:

i)   Between the ends of the frame top bars and the wall of the hive.
ii) Between the self spacing arrangement of one frame and the self spacing
     arrangement of the adjacent frame such as Hoffman frames.

iii) At the contact point on the underside of the frame lugs and the metal runner supporting the frame.

iv) Between the metal runner and the wall of the hive in hives using short lugs such as the Commercial hive.

v) Between the queen excluder and the tops of the frames below it. Slotted types maximise the contact points while the framed Herzog or Waldron types have too much wood that rests on the top bars.

It will be clear that these two sets of criteria have been, by and large, ignored by the equipment designers and manufacturers ever since bee space was propounded by the Rev.L.L.Langstroth in 1853. This is particularly so in the case of frame design and partially so in the case of hive design.

In the following chapters and paragraphs, beekeeping equipment that fails the design criteria will be examined in detail in order to pin point the design faults which can easily be remedied to make colony manipulations easier for both the bees and the beekeeper.

EQUIPMENT THAT FAILS THE DESIGN CRITERIA

Hives.

It is not my intention in this examination of beekeeping equipment to design another hive or even to modify any of the designs presently available. Most of them are satisfactory except for those which are designed for the short lug type of frame and these can be improved to minimise the production of propolis. One exception is the Modified National and the plans show the top bars ¹⁄₁₆ inch below the top of the hive walls in a bottom bee space hive. This gives a bee space of ¼ inch below the frames and can be adjusted during construction to give ⁵⁄₁₆ inch and hence bring the top bars level with the top edges of the walls of the brood box. Similarly, most hives can be constructed for either top or bottom bee space. For more information on hives see 'A case of hives' by Len Heath which is still readily available.

Hive - frame runners.

This is the easiest one to tackle and focuses on one aspect only, that is the metal runners which are preferable to those made of plastic because good bee husbandry requires brood boxes to be cleaned and disinfected from time to time. The easiest way to do this is by scraping and then scorching with a blow lamp. Plastic runners would disintegrate with the heat from the blow lamp. Most metal runners are formed by bending a metal strip to form the ⁵⁄₁₆ inch rebate. See the diagram below. The frame lug lays on the curvature of the bend and forms two spaces which are readily propolised by the bees. There can't be many beekeepers who are not aware of this problem.

To minimise this point of contact the present design of runner should be replaced by a simple strip of metal, say ¹⁄₃₂ inch thick, fixed in the same place to provide ⁵⁄₁₆ inch bee space below the lugs. This will minimize the amount of propolis joining the frame to the runner.

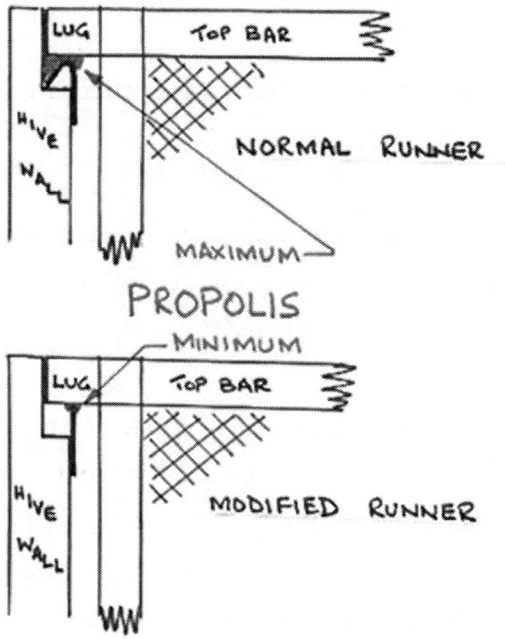

Normal metal runners can be converted at no cost by removing and cleaning them before squeezing them gradually in a vice until the whole length is flattened into a strip. The easiest way to put this back to ensure that the bee space below the frame lug is maintained, is to use a simple jig of wood to correct thickness placed in the rebate. The modified runner can then be positioned just flush with its top edge. Fixing with screws is easier than the usual cheapjack way with gimp pins. Stainless steel ½ inch pan head # 6 are quite suitable.

In the case of hives using frames with a short lug, mentioned earlier, one edge of a metal runner tapers down and abuts the hive wall in the corner of the rebate. The propolis in this crevice quickly joins up with propolis built between the ends of the frames and the wall thereby making the frame very difficult to remove and replace. This is most evident in say the Commercial hive where the rebate is the usual depth of ¹¹⁄₁₆ inch but the lug is only ⅝ inch, or two bee spaces, wide.

Using the simple modification, as shown in the diagrams above, it provides a very effective solution leaving the rebate clean and clear of propolis or comb. It does not overcome the problem of the bees propolising the space between the ends of the top bar and the inner edge of the rebate.

More about this later but here it should be noted that there is no way of avoiding this problem in hives using short lug frames.

Replacing the standard metal runners with flat metal strips should be applied not only to brood boxes but to supers as well in order to comply with the propolising design criterion.

Any standard which is evolved for runners must, in my opinion, specify them being made of metal in order to disinfect the boxes that they are used in by using a blow lamp. Traditionally, metal runners have been made from tin plate (ie. very thin mild steel sheet which is tin plated). There is no reason why they shouldn't be zinc plated (or galvanised) but better still a stainless steel strip, for long and non-corrosive life, would be my preference. It is well to keep in mind that very often brood boxes are used for disinfecting frames using acetic acid.

One final point about metal runners is the inaccuracy of the bee space produced. Over the years before converting to the flat strip type, it was always necessary to adjust the bend of new runners, either more or less, to give the correct ⁵⁄₁₆ inch bee space.

The examination of this simple bit of equipment, used in every hive throughout the land, patently shows the need for a standard specification readily available for all commercial manufacturers. Until such standards are available, the beekeeper will have to put up with sub-standard equipment. This is the tip of the iceberg, there are many other horrors to come.

Queen excluders.

Queen excluders must be the worst example of not complying with the design criterion to minimise the propolising of hive parts. Consider first the slotted types of queen excluders, originally made in zinc sheet with either long or short slots specifically designed to lay across the top surface of the brood chamber top bars. Such an arrangement invites propolisation along every top bar in the brood chamber with the result that the queen excluder has to be peeled off breaking every joint until it becomes free. Touchy bees object to this manipulation. The real problem comes when it has to be replaced; the top bars have to be scraped clean, with more objection from the bees, and the queen excluder itself has to be cleaned. Cleaning the queen excluder is not easy with a hive tool and, if it is of the long slot variety, there is high risk of causing mechanical damage and ruining its excluding capabilities. The only way these slotted queen excluders can be cleaned in the apiary is by carefully scraping them with a hive tool on the top of a flat hive roof. There are mild steel slotted types available which are more robust and there are some made of plastic. All fail the design criterion for minimising propolisation and should be ignored and abandoned. There is no way that they could be adapted to be even partially successful. They are cheap but it is a false economy, in my opinion, to purchase them.

The next type (wire) include the Herzog and Waldron variety; both are similar in construction with a robust metal wire grid enclosed in a wooden frame. A queen excluder for a National hive will be discussed but the principle is the same for all hives. The wooden frame is made from timber finished to ³⁄₈ inch thick or, more common these days, timber finished to 9 mm. The wire grid should be flush with one side of the wooden frame with a bee space on the other side. Sometimes there is and sometimes there isn't, depending who made the frame. Don't get caught when you buy a new one.

The wooden framing of queen excluders vary in width, again depending on who made them, and one that was measured from a stack of 25 in my apiary shed had 2 sides 1¼ inches wide and the other 2 sides 1⅛ inches wide with a rabbet on one side to take the wire grid. The rabbet was too deep leaving a gap of approximately ⅛ inch at each side which is filled with propolis. The frame is jointed at the corners but it is unglued and stapled together cheapjack fashion. The frame should ideally be of the same profile as the cross section of the hive it is to fit; in the case of a National, 2 sides should be ½ inch wide and the other 2 sides ¾ inch wide. Study of the diagram below will make it abundantly clear that the woodwork will be in contact with all the ends of the upper surfaces of the top bars and one of the top bars throughout its whole length. A most unsatisfactory situation. While the arrangement is much better than the slotted types it still fails the propolis design criterion. Prising the wooden frame from the top bars at every manipulation soon weakens the joints and most of mine have been remade and glued.

So what is the solution? In my opinion it would not be possible to make a wire type excluder with a wooden frame with sides of ½ inch and ¾ inch width, it would be insufficiently robust always assuming that the grid is large enough to fill a frame with larger inside dimensions.

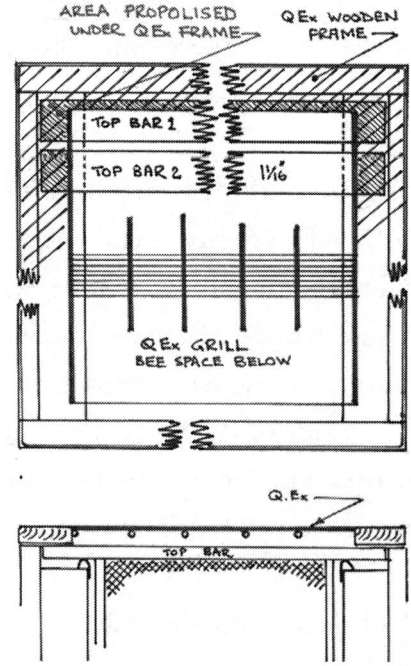

I believe that the frame should be made of mild steel to the correct dimensions of the hive it is to fit with the grid welded into it and the whole zinc plated after fabrication. I discussed this matter with an engineering firm making and selling wire type queen excluders on a visit to Stoneleigh. He thought that I was 'winding him up'. I assured him that I wasn't winding him up and he then wanted to know how many I would purchase. Different hives have different wall thicknesses and it would be necessary to specify the hive it is to be used on; this should not present a problem and a common specification could be written which would embrace all types of hive.

Metal/plastic ends and other spacers.

There is quite an array of spacing devices available from the equipment suppliers for spacing frames both in the brood chamber and in the supers. Each will be examined to evaluate its efficacy.

a) Metal ends.

Metal ends have been around for a long time and originated, I know not when, but with them 10 frames just fitted snugly into a WBC brood box taking up a width of 15½ inches. The metal ends were made not 1½ inches wide but $1\frac{9}{20}$ inches. The spare ½ inch ($\frac{1}{20}$ x 10) was very quickly taken up with propolis and the end frames were very difficult remove and replace. Metal ends of this size are known as narrow metal ends while there is a wide metal end for super frames which gives a spacing of 1⅞ inches. Propolis makes the removal of metal ends a hazardous occupation (sharp edges) at extracting time in order to fit the frame into the extractor and cleaning them for re-use is a thankless chore. Happily, they can banished from the beekeeping scene for ever without any loss to the craft.

b) Plastic ends normal and 'V' type.

The normal plastic end suffers from the same old problem of having a large surfaces abutting each other to provide the 1½ inches frame spacing causing propolisation problems. These have been superseded by the 'V' type which has 'V's at both ends with one 'V' horizontal and the other 'V' vertical which then gives a point contact when the frames are closed up - excellent, the first and only bit of equipment that has had some thought involved in its design as far as bee space is concerned. They could be purchased in a variety of colours which could be used to denote the year the foundation was fitted and hence it was readily possible to know the age of the comb in each brood chamber. However, their disadvantages outweigh their advantages as they suffer from most of the metal end problems, they too can be banished from the beekeeping scene.

c) Castellated spacers.

These do meet the criterion of minimising the contact with the top bars as the lugs are sitting on a virtual knife edge. The only criticism would be that the vertical slots are unnecessarily deep and an additional two edges are provided for the bees to propolis. They originated for use in supers to get over the problem of having to remove all the metal ends when extracting. How ever, they may be purchased for 9, 10 or 11 frames in a National hive and to take 8 or 10 frames in a WBC hives. There are better ways of 'running a railway' than using these devices and another horror can be relegated to the junk bin.

d) Yorkshire spacers.

This spacer was designed to clamp on to the side bars of a frame giving 1⁷⁄₁₆ inches spacing. It is made of tin plate like a metal end and suffers from the same defects. It can safely die a natural death.

It will now be clear that no specification is needed for separate spacers and all frame spacing both in the brood chamber and in supers is best effected with self spacing frames a topic of major importance dealt with in the following paragraphs. A point of interest arises from this study of spacers and that is why there should be three different spacings for use on the same frames in the brood chamber with the same bees? Answers on a post card please.

Frames general.

A brief look at any of the equipment manufacturers' catalogues will show a bewildering array of available frames. These are shown in the table :

Designation	Top bar	Side bars	Bottom bars
DN1 & SN1	⅞" wedge	⅞" plain	2
DN2 & SN2	1¹⁄₁₆" wedge	⅞" plain	2
DN3 & SN3	⅞" saw cut	⅞" plain	2
DN4 & SN4	⅞" wedge	1⅜" Hoffman	2
DN5 & SN5	1¹⁄₁₆" wedge	1⅜" Hoffman	2
DN6 & SN6	?	?	?
SN7	1¹⁄₁₆" wedge	1⅜" Hoffman	2

Notes and Key to abbreviations:
 D = deep (8½ inches for a National)
 S = shallow (5½ inches for a National)

The DN3 and SN3 saw cut top bars are very unusual these days probably due to two reasons. Firstly, they harbour wax moth eggs and secondly, they are very difficult to fit the foundation in the saw cut.

DN6 and SN6 have never been allocated to my knowledge and it is a bit of a mystery why 7 was used and not 6.

It is possible to obtain a solid bottom bar all in one piece in lieu of the two bottom bars but, again these are a rarity these days because it makes the fitting of foundation that much more difficult. All plain side bars have a slot on the inner side as a guide for the wax foundation. The plain means that there is no self spacing arrangement. Each DN and SN designation can be applied to any hive so when these are multiplied up by the hives available the array is quite formidable. For example, in the UK the following 5 hives are in use: WBC, National, Commercial, Langstroth, Smith.

All those frames with plain side bars require the addition of a pair of spacers, one at each end of the top bar. It will be recalled that none of the available spacers complied with the criterion for minimum propolis, therefore, only DN/SN 4&5 together with SN7 remain to be examined in order to decide their efficacy. These brood and super frames will be considered in the following paragraphs.

Brood frames.

If the combs in a feral colony are examined carefully it will be found that the comb spacing, septum to septum, is exactly 1⅜ inches (35 mm) and the comb thickness is ⅞ inch (22 mm) making each brood cell ⁷⁄₁₆ inch (11 mm) deep. The space between comb faces is readily seen to be ½ inch which is two bee spaces of ¼ inch each, allowing the bees to work back to back on adjacent comb faces. This is illustrated in the diagram below where comb has been built downwards on a flat horizontal surface. The extremities of each comb form a curve known as a catenary curve. If each comb face is examined the extremity of the comb is also in the shape of a catenary.

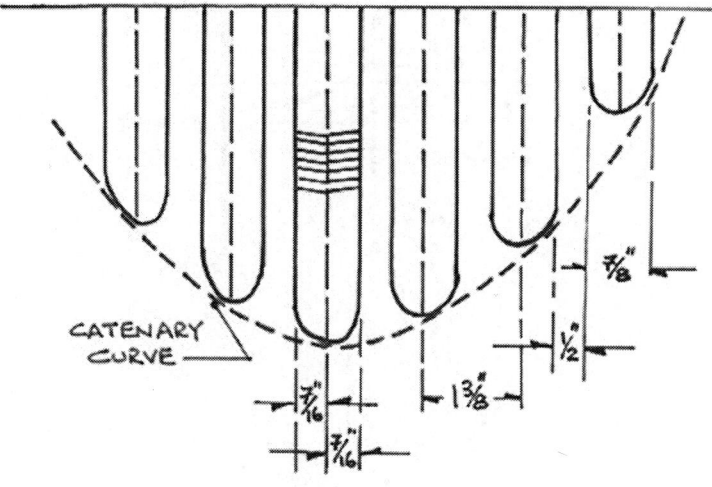

Wild Comb
X-Section & dimensions

The only frames available commercially that satisfies the comb spacing criteria are those with Hoffman spacing built into to the side bars (ie. with a 'V' shape on one side abutting a flat on the other side). It seems very curious that only Hoffman managed to get the septum to septum distance right. Metal and plastic ends with spacings of 1⁹⁄₂₀ inches and 1½ inches respectively do not satisfy the spacing criteria and it is not proposed to discuss them any further. Similarly, Yorkshire spacers with a spacing of 1⁷⁄₁₆ inches need no further consideration.

If a frame spacing of 1⅜ inches is adopted the only top bars that can be used with such spacing are those with a width of ¹¹⁄₁₆ inches which then gives a space of ⁵⁄₁₆ inches between the top bars which is the old familiar bee space. Equipment manufacturers make top bars with a width of ⅞ inch which is a hang over from the days when metal ends were popular. These top bars give a space between top bars of ½ inch which is too wide and the bees will build brace comb between them making manipulation difficult. Avoid them like the plague and the commercial suppliers will soon stop stocking them.

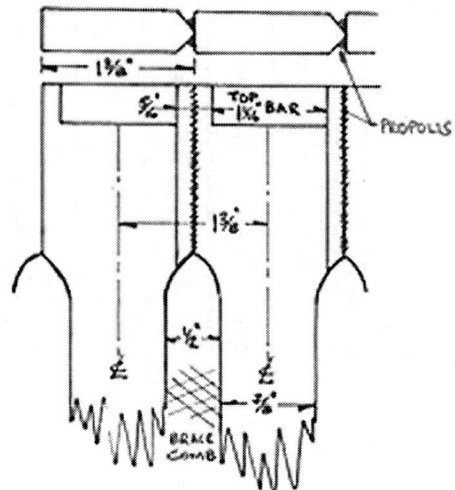

Hoffman spacing

While Hoffman got the inter frame spacing correct, he failed to observe the criterion of minimising the surfaces which become propolised, ie. the couple of inches of contact between the spacers on the shoulders of the frame (see the diagram above). This makes it necessary to lever the frames apart with a hive tool as the propolis builds up down both sides of the 'V' abutting the flat on the adjacent frame. To minimise this problem, the only solution is to plane off the 'V' and replace it with a small round headed brass or stainless steel screw (¾ inch # 6). The only place which will then be propolised is around the screw head which makes a point contact with the adjacent frame. This minor modification could easily be arranged by the manufacturers and at the same time reducing their manufacturing costs by not having to form the 'V' shape. Such a frame would then meet the criterion of minimum propolis associated with the spacing device.

The next failure is associated with the side bars which are usually ⅞ inch wide which gives a space of ½ inch between adjacent side bars which we have already seen is unacceptable. There is not much that can be done about this without recourse to a radical redesign of a suitable brood frame.

It will now be becoming clear, in the list shown previously on page 19, that only DN5 partially meets the design criteria, all the rest fail abysmally. In fact, study of the classical beekeeping literature during the last century reveals that virtually no attention has been paid to the design of frames and spacing to comply with the known paradigm.

The penultimate failure of all brood frames is that the top bars are always propolised to the walls of the hive at their ends. The inside width of the hive is always $\frac{1}{16}$ inch greater than the length of the top bar allowing a tolerance of $\frac{1}{32}$ inch at either end. This is good engineering practice allowing this end play but unfortunately it is just the sort of gap that the bees propolise and thus the frame fails the propolis criterion. The matter is simply rectified by cutting off $\frac{5}{16}$ inch from both ends of the top bar and replacing the missing wood with round headed screws (1 inch # 6) adjusted to give the original length of the top bar which still gives the tolerance and invites propolising but it now minimised to around the screw head.

It is interesting to note that over a century ago frames were made in some hives cut to a point at their ends, no doubt to minimise propolising but for over a century 99% of all frames exhibit this problem.

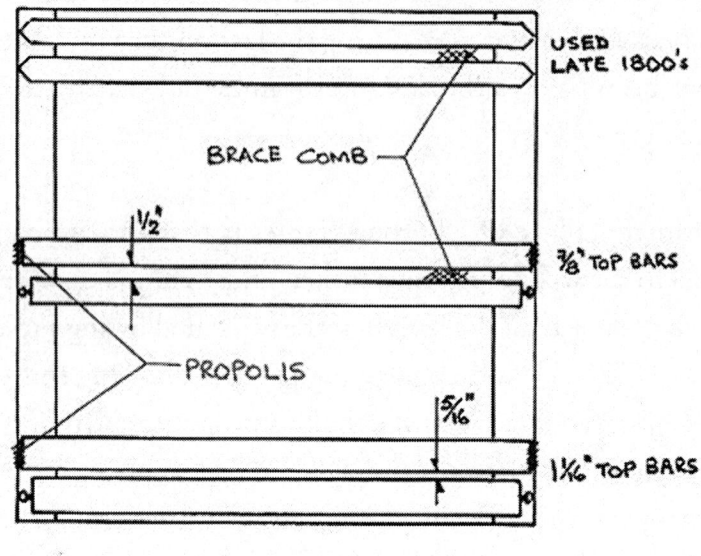

Top bars

The final failure, again with all brood frames, is the slot between the bottom bars which invites immediate propolisation. A solution using commercially produced frames is not obvious and the solution would require redesigning a new frame to provide a slot of $5/16$ inch complying with the bee space criterion and making it infinitely easier to clean out and replace with foundation when required.

So what is the answer to this problem of frames and spacing failing such fundamental requirements? It beggars belief that the problem of propolis, at the ends of the top bars, was recognised in the late 1800's and has been forgotten for over a century. While it is possible to modify Hoffman frames and effect a very noticeable improvement the real answer is to design a frame that complies with and one that fits existing hives. Luckily this is possible and the solution that I offer would make for a much cheaper frame with minimum machining to manufacture the parts. It would require to be manufactured in two versions, the first for a wired frame and the second for use with wired foundation with a wedge type top bar. It is proposed to describe the former which would be the simplest and the cheapest to manufacture; it also has the advantage of keeping the foundation flat in one plane rather than using wired foundation that often warps.

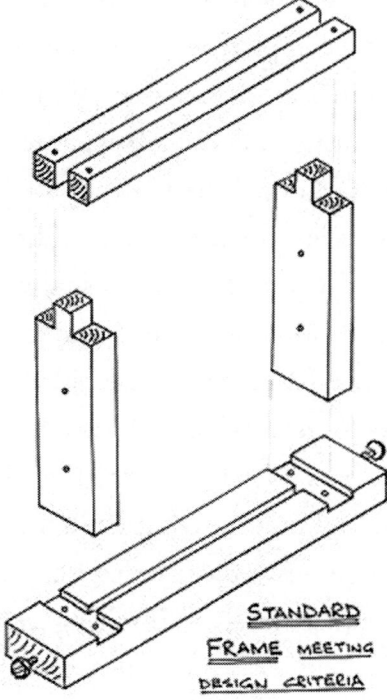

STANDARD
FRAME MEETING
DESIGN CRITERIA

The warping is due to the expansion or contraction of the wire which is inevitable if the ambient temperature is above or below the temperature at which it was made.

In order to make the frame as simple as possible for manufacturing purposes, both commercially and for construction at home the timber thickness is the same for all parts, that is, ⅜ inch thick from any knot free timber which is unlikely to warp. The parts required are as follows for say a National hive:

1 -Top bar: ⅜ inch x 1 ¹⁄₁₆ inches wide x 16⅜ inches long

2 - Side bars: ⅜ inch x 1 ¹⁄₁₆ inches wide x 8³⁄₁₆ inches long

2 – Bottom bars: ⅜ inch x ⅜ inches wide x 14 inches long

These can be produced on a circular saw with fine teeth and need not be planed. If the saw is fine enough and the speed of rotation high, the finish will be perfectly adequate for hive furniture and thereby saving a planing operation in the manufacturing process. There are two $\frac{1}{16}$ inch slots to be routed in the top bar to position the side bars and a central slot to locate the foundation on the underside. Two $\frac{3}{8}$ inch x $\frac{3}{8}$ inch cut outs in each side bar are needed to receive the bottom bars. This completes the actual machining leaving only 8 -1.5 mm holes to be drilled as guides for the copper nails to assemble the frame with a waterproof glue. The diagram above shows an exploded view of the simple frame. Further 1.5 mm guide holes can be drilled in the side bars for the final enlarged drilling to accept the brass eyelets for wiring. An additional 4 guide holes should be drilled to take the # 6 stainless steel pan head screws for frame spacing and for the top bar ends. All these drilling operations may be performed commercially and at home by the use of jigs. The frame is assembled with $\frac{3}{32}$ x $1\frac{1}{4}$ inch copper nails to attach the side bars to the top bar and $\frac{1}{16}$ x $\frac{3}{4}$ inch copper nails to attach the bottom bars to the side bars. Needless to say that the finished frame must be square and free from skew. Holding the completed frame by its lugs with the top bar upper-most, the spacing screws must be on the RH side bar when viewed from both sides. To set these accurately, a simple jig should be made from a scrap piece of timber with a $1\frac{3}{8}$ inch cut out on one edge for measuring the distance the screw should be inserted. Clearly if the side bars are $1\frac{1}{16}$ inches wide the screw will protrude $\frac{5}{16}$ inch for $1\frac{3}{8}$ inch spacing.

A similar frame to accommodate wired foundation will have a modified top bar to form the wedge instead of the slot on the underside to locate the wax foundation. It will have a smaller area due to the space required to form the wedge. The arrangement is so familiar that a description will be omitted here.

As there is only one spacing screw on the RH side of the frame, the first frame adjacent to the hive wall when pushed up tight will not be parallel to the hive wall. A spacing screw needs to be inserted in the brood box wall protruding $\frac{5}{16}$ inch to square this up. An identical arrangement is needed in the opposite hive wall as well as on one side of the dummy board.

This completes the description of my frame which is offered as a contender for a standard frame. There may be better ideas but this one meets all the criteria for bee space and propolis and has the added advantage of simplicity and cheapness. If assembled with glue and copper fastenings with stainless steel spacers it will withstand disinfection with acetic acid without the usual resulting nail sickness when steel gimp (a word that doesn't seem to make sense when used in this context!) pins are used. For migratory beekeeping it is recommended that two spacing screws are used on each side of the frame, one at the top of the side bars and one at the bottom in order to prevent them swinging during transportation. If my recommendations are observed, particularly in respect of the frames and runners, there will be a minimum of propolis to gum the frames up which is usually an advantage when colonies are being moved. In fact, it is the only advantage of propolis for practical beekeeping purposes.

It should be noted that screws can only be used at the ends of the top bars in hives that use frames with long lugs. Those hives with short lugs (⅝ inches) cannot have 5/16 inch cut off each end. Perhaps there is merit in going back 100 years and tapering the ends of these short lug top bars (see the diagram on page 24).

Super frames

There are basically two types of super frame, those spaced with metal or plastic ends and the self spacing Manley frames. Manley, many years ago, experimented with super frames to find the widest spacing that the bees would tolerate when presented with supers of foundation so that it was drawn out to form evenly constructed combs. His verdict was a spacing of 1⅝ inches; this spacing has never been disputed and it has stood the test of time. Most beekeepers when purchasing super frames now buy Manley, which are self spacing by using side bars with a width of 1⅝ inches. These side bars abut one another throughout the super and there is no space between the side bars. The maximum width of top bar available is 1 1/16 inches making the space between adjacent top bars equal to ⁹⁄₁₆ inch which again fails the bee space criterion and which can be bridged with brace comb.

Other frames using metal and plastic ends come in 1½ inches, 1¾ inches and 1⅞ inches spacing which would fail all round on the bee space criterion. Try now calculating the space between top bars using a top bar width of ⅞ inch for Manley spacing of 1⅝ inches.

Anyone who has used Manley frames in their supers will have noticed that the area for propolising has been maximised in the design, not minimised, and the bees readily propolise all the frames together. If they are not levered up hard against one end of the super it doesn't take long for them to fill up the whole available space and make it very difficult to get the frames in and out. The answer to this problem is again very simple; cut off ⁹⁄₃₂ inch from one side of each side bar and replace this with a round headed screw (¾ inch # 6) which again minimises the area for propolising to that around the point contact of the round head part of the screw. The reader is left to reason why the amount to be cut off is ⁹⁄₃₂ inch when using a 1¹⁄₁₆ inches top bar.

Apart from the propolisation aspect with the side bars, Manley frames have been considered to be eminently suitable because the bottom bars have the same width as the top bar, ie. 1¹⁄₁₆ inches thereby providing guides for uncapping and leaving parallel sided comb after extraction. It is also of a thickness very suitable for cut comb which, of course, was Manley's objective when he experimented on wide spacing. We must ask ourselves whether the Manley frame with the slight modification to minimise the propolisation problem is the best that we can do? I am not sure about this as will become apparent in the following paragraphs.

It will be abundantly clear from the foregoing discussion on brood frames that if a given spacing is required with a bee space of ⁵⁄₁₆ inch then the frame must be constructed with top bar, side bars and bottom bars all of the same width (ie. frame spacing minus ⁵⁄₁₆ inch). Similarly, should the same reasoning applies to super frames? If we wish to make a super frame with Manley spacing

(1⅝ inches) then it must be constructed in timber 1⁵⁄₁₆ inches wide.

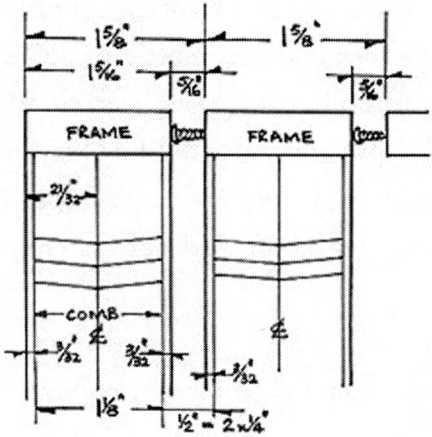

Let us ponder such a frame for a moment or two and consider its pros and cons to decide whether there is a worthwhile improvement over what is available. Again we will consider a National super but the reasoning will apply to all other hives. With reference to the diagram above, it will be clear that if the bees have two bee spaces each of ¼ inch between the comb faces, then the comb face will ³⁄₃₂ inch below the edge of the woodwork if top bar, side bars and bottom bars are all the same width (1⁵⁄₁₆ inches). This would preclude using the top and bottom bars as a guide for the uncapping knife at extraction time. This argument is true for frames and comb of any thickness; nb. a frame spacing of 1³⁄₈ inches with woodwork 1¹⁄₁₆ inches wide and a comb thickness of ⅞ inch, the comb face will also be ³⁄₃₂ inch below the woodwork of the frame; however, this is a typical brood frame but it follows the same pattern when a bee space of ⁵⁄₁₆ inch is provided between the woodwork of adjacent frames. We must conclude, reluctantly, that what appeared to be an elegant and simple solution for brood frames cannot be applied to supers for extraction. It appears that if the top bar and the bottom bars are to be used as a guide for the uncapping knife then the width of the timber must be reduced thereby increasing the space between adjacent top bars and bottom bars to something greater than a bee space and thus encouraging the building of brace comb.

Years ago when top and bottom bars of ⅞ inch were 'de rigour', all the classical literature talked about uncapping by running the knife just below the cappings. If the woodwork had been used as a guide for the uncapping knife an unnecessary amount of comb would have been cut off in the process. It would be most desirable if a few other minds could be bent around this problem in a search for an elegant solution. In my own case I use Manley frames with 1¹⁄₁₆ inches top and bottom bars with modified side bars to prevent the propolisation problem.

The top bars of super frames require to be shortened by cutting off ⁵⁄₁₆ inch from each end and replacing it with a screw to minimise the propolising of the ends to the hive wall within the rebate for the lugs.

One final point about supers which is seldom addressed in the classical literature and that is about cracking supers. In the old days it was always recommended that supers were cracked before removing them for extraction. By this it was meant that each super was removed one at a time and all the brace and burr comb cut off and removed. It is a sticky messy job but well worth the trouble when it comes to removing them a day or two days later. They have all been dried up by the bees. It therefore makes handling them between the apiary and the extracting room a clean job and honey is not dripping from the frames in the extracting room while the frames are being processed.

This is a hang over from the days when most frames were made up from narrow top and side bars with wide spacing, thereby encouraging the building of brace and burr comb. Even with modified Manley frames in a good flow the bees still tend to build excess wax comb.

We have always cracked our supers and have a bucket of water with us for washing the hive tool and our hands after completing one hive before going on to the next one.

OTHER EQUIPMENT REQUIRING A RE-THINK

Smoker and hive tools

A smoker and hive tool are the two primary tools used by all beekeepers in order to manipulate their colonies. They should never fail but, unfortunately, the average copper smoker is a shoddy bit of work and usually fails at the wrong time, as will be explained later, while the hive tool is very reliable but very expensive and doesn't fit snugly in the pocket of a bee jacket or overall because it is too long. Both items are addressed in the paragraphs below.

Hive tools

There are basically two types of hive tool available from the equipment suppliers both made from stainless steel about 9 to10 inches long. The traditional type has the flat blade at one end for splitting boxes with its other end bent round to act as a scraper and lever. The other type, often referred to as a 'J' type has the flat blade and the other end is shaped to act as a frame lifter. This 'J' type can be purchased in mild steel and is often painted red.

I have never understood why anyone, in this day and age, should want a 'J' type with the frame lifting gizmo, as frames are generally moved in a horizontal direction with a hive tool when manipulating a colony. I believe that they originated in the days when just the right complement of frames filled the brood chamber and the dummy board had not yet been brought into general use. It was then very difficult to remove the first frame for inspection. Every enlightened beekeeper today uses one less frame in the brood chamber and a dummy board making the frame lifter redundant. They are marginally cheaper than the traditional type and could be dispensed with; beekeeping would not suffer if another was not manufactured.

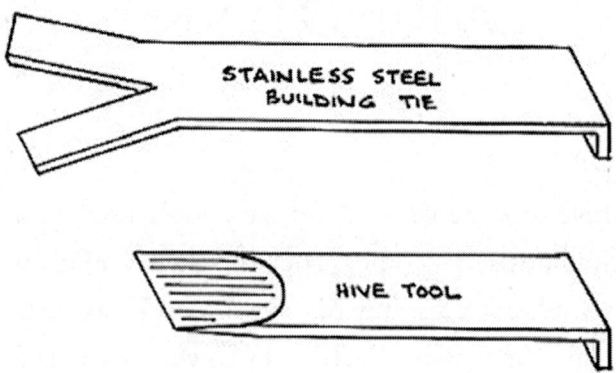

The traditional type, costing about £10, is too long, in my opinion, and I prefer one which fits in my jacket pocket when not being used so that I do not lose it. I have a standard number of items in my jacket pockets which include a disposable cigarette lighter for lighting the smoker, a Butler cage with wooden plug for caging a queen at short notice, a sharp penknife for cutting out queen cells when required, my own design of hive tool and a pen for filling in records. My hive tool is short (5 inches overall) and costs 50p. It is made from a stainless steel building tie which is readily available from most good quality building merchants. Cut off the end and grind a flat blade using a grind stone and then put your name on it! If frames and hive parts are designed for minimum propolisation then a long hive tool is unnecessary as minimum leverage is required. The hive tool is shown in the diagram above. To ensure that you get a stainless steel building tie, take a little magnet along with you to check; s/s is non-magnetic.

Copper smoker

Generally, copper smokers are marginally cheaper than those made from stainless steel, the latter could be better value for money. However, some of the comments that follow also apply to the stainless steel variety. Since by far the greater number of beekeepers own the copper variety and they always fail (the smoker that is!), my examination of this type is made for two reasons, namely:

To provide the detail on how to rehabilitate one that has broken down.

To provoke some thought for those who manufacture them in the hope that something more workmanlike will be made in the future.

The diagram below shows at upper left a typical copper smoker which has been made in this form for many years. The parts that cause failure are marked and each is discussed below:

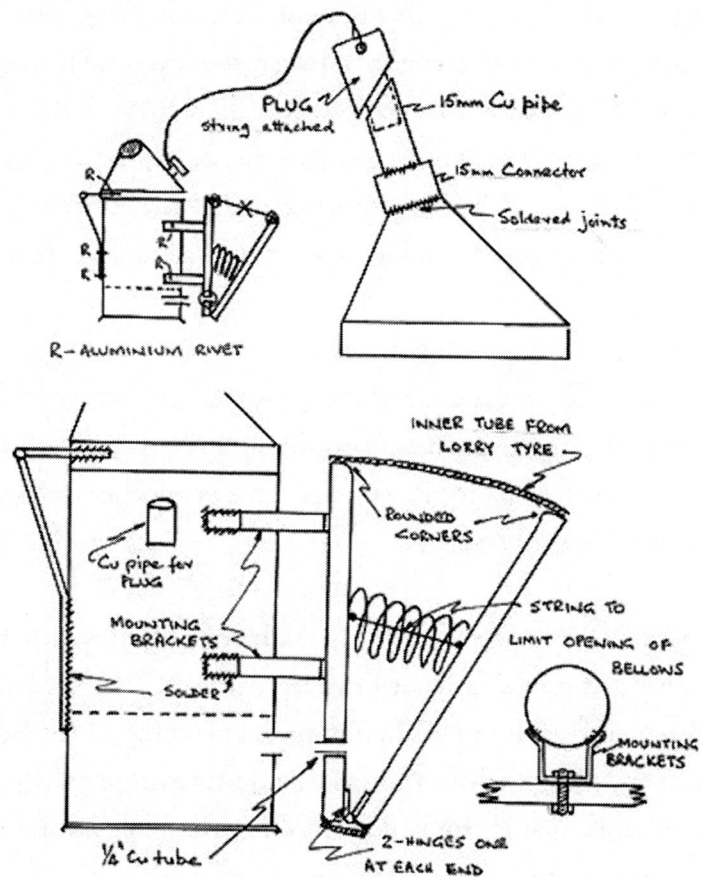

1. The hole in the lid of the smoker where the smoke comes out is too large and when the smoker is lit and burning well it continues to burn on its own when stood in the upright position. The smoker should continue to burn without operating the bellows but it burns too quickly and it is necessary to keep adding fuel at short intervals. It is possible to lay the smoker on its side to prevent the self-chimney effect but then there is a chance that it could go out in the middle

of a manipulation. A much better solution is to reduce the size of the hole which reduces the chimney effect and it provides a more direct puff of smoke when used.

2. The smoker is assembled with aluminium rivets using a pop riveter. Over a period of time the rivets disintegrate and the lid and the bellows will fall off. When combustion takes place water is produced (have a look at the amount of water coming out of a car exhaust on a cold morning just after it has been started). The water acts as an electrolyte between two dissimilar metals and one is 'eaten' away. It is not necessary to go into the chemistry of this reaction but it occurs on most boats and unless a sacrificial anode (usually made of zinc) is fitted, it is not long before the bronze propellor drops off. In the smoker the soft aluminium rivet is the sacrificial anode. This is a basic design fault and needs rectification.

3. The air inlet/outlet hole in the bellows after a few seasons use becomes charred and will slowly enlarge due to the heat from the firebox directly opposite and in line with the hole in the firebox in the smoker. This is another design fault albeit the least serious of all the others.

4. The smoker bellows are made of rexine (a trade name for a synthetic imitation leather), a cheap non-durable form of flexible covering. The large spring in the smoker keeps the rexine under tension. Continual working of the bellows wears away the rexine at the points where it creases and more importantly at the sharp edges of the woodwork inside the bellows. When the bellows start to leak the repair is usually a bit of sticky tape and then another bit until it is time to fit a new bellows and guess what, the equipment suppliers provide the same rubbishy spare made in rexine! The other cheap and nasty feature is using the rexine at the bottom of the bellows as a hinge. There are no proper hinges inside the bellows.

5. The final fault is that there is no permanent plug attached as an integral part of the smoker and it is necessary to stuff the hole with a wad of green grass to extinguish it after use. Many fires have been caused by emptying the hot ashes

from the smoker after use; never ever do this. Put a plug in the smoker and leave it in the upright position to go out and cool down of its own accord.

These are the faults and every beekeeper who owns such a smoker must be well aware of them. The manufacturers should be encouraged to provide a better designed tool but meanwhile the show must go on and the following tips will turn your smoker into something that you can rely on. Taken in the same order as listed above:

1. This is very easy to fix with a bit of 15mm diameter copper tube used for household plumbing and central heating systems and a 15mm/19mm reducer all soft soldered into the smoker outlet. The hardest part of this part of the job is cleaning up the old copper surface for soldering. It should be done with wire wool and the surface to be soldered must be absolutely immaculate before applying the flux to both parts before soldering.

2. The brackets attaching the main body of the smoker to the bellows are made of aluminium. Remove them and make two new ones from flattened ¼inch (8 mm) copper tube. The part of the bracket that is to be soldered to the main body of the smoker will cover the holes left by the old rivets. The other soldering job is on the smoker lid hinge which is made of mild steel and will be very difficult to clean before soldering. When the underside of the hinge is clean it is easier to pre-tin this with solder before actually fixing it to the smoker body by soldering. There is an alternative to soldering and that is pop rivetting but this time using copper pop rivets. There will be no electrolytic action as everything is now copper, the smoker body, the newly made brackets and the fastenings, the rivets. I had difficulty finding the copper rivets but eventually located them on the web at www.rivetwise.co.uk; they are based in Birmingham on telephone number 0121 766 5445 and small quantities can be purchased direct from them. It should be noted that copper rivets can be purchased with an assortment of mandrels made of carbon steel, bronze or stainless steel. Only stainless steel should be used to obviate electrolytic action.

3. When the bellows have been disassembled fix a short piece of copper pipe in the inlet/outlet hole; a bit of ¼ inch copper tube has been a drive fit in the smokers that I have fixed.

4. Now comes the major rehabilitation of the bellows which will make it infinitely easier to re-assemble. First fix two small brass hinges at the base of the wooden sides to join them together after removing the sharp edges with sand paper. The next operation is to fix an attachment to each wooden side piece to act as an anchoring point for a piece of string or terylene cord to limit the outward travel caused by the spring. The bellows are now ready for covering. I have tried various materials but the best so far (and I suspect will see my time out) is a bit from an old discarded heavy duty inner tube from a lorry. Easily obtained for the asking at most garages that change tyres. Your finished bellows should not be under any strain, the string inside is taking the tension. The bellows will last a lifetime so do a good job on the string inside!

5. Finally, solder a piece of 15 mm copper tube onto the side of the smoker to house the wooden plug on a string when the smoker is in use.

This rehabilitation is a major overhaul job best done in the winter months when the smoker is not required. It would be better if the original manufacturer could get it right in the first place; it would cost little more and a specification would be so easy to write for them to work to.

Floorboards

The only thing that seems to have changed on the average floorboard is the floor itself. They were originally planked with ¾ inch thick timber and the tongue and grooved joints were a haven for wax moth. Plywood has replaced the planking and the floor thickness has become thinner with ½ inch ply usually being used. Sometimes the plywood warps slightly and the entrance block does not fit snugly with the floor throughout its length. This warping was discovered moving some stocks of bees. On the journey, bees were escaping either side of the slot in the entrance block. The 4 inch entrance slot had been plugged with polythene foam but the bees found the gap at the sides of the entrance slot where the floorboard had warped; pushing the foam in tightly is likely to make the gap bigger.

The remedy is simple. Glue and screw a piece of hard wood on the under side of the floorboard at the front to stiffen it.

Entrance blocks/mouse guards

Years ago, while experimenting with wintering, we hit on the idea of incorporating the mouse guard in the entrance block. It was so successful that we have used it ever since and that must be some 30 years ago. Our method of wintering is to cover the feed hole in the crown board, raise the crown board with four matchsticks, one at each corner and insert our special entrance block.

The entrance block has 9 holes drilled in it each of ⅜ inch diameter as shown in the diagram. The area of the 9 holes is 1 square inch and this limits the ventilation through the hive in winter.

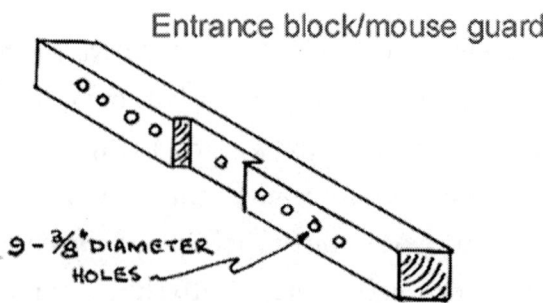

Entrance block/mouse guard

9 - ⅜" DIAMETER HOLES

About four gallons of water have to be disposed of during the winter months and it is necessary to raise the back of the hive about ¾ inch to drain some of this water out at the front; failure to do this will result in a very wet floor board in the spring. The other feature of this entrance block is that it makes a very efficient mouse guard; no mouse can enter and bees can pass freely backwards and forwards particularly in the spring when collecting pollen. The pollen loads are not knocked off as they enter the hive as happens with those horrid metal types which are attached with drawing pins. Note that a mouse can enter through a ⅜ inch slot but not a ⅜ inch diameter hole because its skull is wider than ⅜ inch.

Before leaving this topic, a word of warning about the hole size. Make it exactly ⅜ inch diameter as specified; one not so bright beekeeper copied our idea but drilled holes of ³⁄₁₆ inch diameter with the result that he killed off his 5 colonies as the bees could not get out. It was a very sad sight.
The entrance block cum mouse guard should be made of hard wood (oak, mahogany, etc.) rather than soft wood. The bees will chew around the holes and a soft wood version will only last 2 or 3 winters at the most.

All entrance blocks should be stored in the roof above the supers when not in use and should be flamed every year with its floorboard when changed in the spring. Storing the entrance block in the roof may create a problem which is addressed under the heading roofs in the following section.

Roofs

There are one or two problems with roofs as supplied by some of the manufacturers which definitely require addressing.

Roofs for most hives come in three depths, namely, shallow (4 inches), medium (6 inches) and deep (9 inches) and it is not clear how these came about. The deep roof of 9 inches is not very common these days and it is usually the 4 inch and 6 inch roof depths that appear in the catalogues. The deeper the roof the more difficult it is for the wind to lift it and blow it off the hive in inclement weather. The clearance between the outside of the hive wall and the inside of the hive roof should be not greater than $5/16$ inch. A 6 inch roof with the correct clearance is very unlikely to blow off; I have never had one blow off and mine are never weighted down with stones or concrete blocks. I have no experience with the 4 inch variety which were not originally specified by the Ministry when they produced the drawings for the Modified National and the Commercial hives. The major problem are those manufacturers that produce roofs with too large a clearance and I have seen some which are truly ridiculously large. Avoid them as they will give trouble sooner or later.

The next point are the battens fixed around the inside of the roof to give a head space of $1\frac{1}{4}$ inches above the crown board. This head space is fine for storing the entrance block, when not in use, diagonally across the crown board. Unfortunately in order to cut costs the battens have been pruned down in size and very often cannot accommodate the normal sized entrance block of $7/8$ inch square. It is simple to rectify by glueing and screwing some additional laths of wood to the existing battens to get the correct head space.

The non-rusting metal is usually put on with galvanised iron clouts which in time work out and corrode. A far better arrangement is to fix the metal on with stainless steel screws. Even zinc plated screws eventually lose their plating and rusting occurs. In high summer the temperature of the metal sheeting becomes so hot that it is impossible to put a hand on it for any length of time.

To reduce the heat in the head space under these conditions a ½ inch thickness of polystyrene sheet beneath the metal sheet helps considerably to make life tolerable at the top of the hive. Lift the roof in summer when the colony has 3 or 4 supers on and bees will be seen around the feed hole fanning to remove hot air and moisture from ripening honey in the supers. It continually puzzles me why many beekeepers either close the feed hole completely or cover it with zinc gauze; consider the amount of work the bees have to do to push all that air down the hive and out at the front.

When manipulating colonies in the summer, normal procedure is to place the roof upside down at the rear of the hive with a corner facing the rear wall. The supers are lifted off and placed on the upturned roof which has to take a considerable load of up to about 150 pounds. In time the joints at the corners of the roof become a bit tired and start to work loose. This can be prevented by ensuring that no movement takes place by nailing or screwing 'L' shaped metal pieces to the underside of the roof corners. All mine are made in sheet copper. These corner pieces were a standard feature years ago but have disappeared as a result of cost cutting.

The final point about roofs are the ventilators. The standard arrangement is to cover the ventilators on the inside with a piece of gauze but the gauze that is generally supplied is too fine and impedes the ventilation. Gauze of about 12 per inch in 26 standard wire gauge is suitable. The ventilator soon fills up with spiders webs and cocoons which is difficult to clear; the insects responsible enter from the outside. A further gauze should be fitted to the outside for two reasons:

1. To keep the ventilator clear as explained above, and,
2. To prevent silent robbing which occurs quite often and has largely been ignored in the classical literature on honeybee husbandry.

It will be apparent that the simple roof has quite a few things about it that need to be adequately specified for the manufacturers to work to. There may be a more elegant solution to the ventilator problem and consideration could be given to say a fibre glass cover instead of metal. Such a cover could be easily manufactured after a suitable mould has been made. Alternatively, a complete roof in fibre glass may be a better solution.

Crown (Cover) boards

These simple boards suffer from the same fault as the queen excluders and require the fillets on the underside, for a bottom bee space hive to match the thickness of the hive walls where they mate with the crown board, whether it be a brood chamber or super. If they are made in such a fashion then it would be important to orientate them correctly on the hive.

It is not necessary to provide two holes for Porter bee escapes, 1¼ inch to 2 inch diameter holes are perfectly adequate and easier to manufacture. Clearer boards are a separate item and crown boards should not double up as clearer boards. The reason is simple; if the crown board is used as a clearer board the colony will then have no crown board and the roof will be in direct contact with the super thus putting propolis on the roof battens. Some time in the future the roof will stick to the crown board and will be difficult or very difficult to remove.

Clearer boards.

There are a variety of different types of clearing devices that fit into a simple board. Each beekeeper will have his own favourite method but the board remains substantially the same as far as designing for bee space and minimum propolis. Again, the fillets should match the wall thickness of the hive in use and an asymmetrical arrangement will result requiring the correct orientation to be observed when in use.

Feeders

It is considered that only one type of feeder is required and one per hive is required because once feeding starts in the late summer all stocks in the apiary should be fed together. Anyone needing to feed their bees in the spring should not be keeping bees; it means that the colony was not properly prepared for winter the previous year.

The choice comes down to the Miller or the Ashforth, both fast feeders, and in both cases the fillets on the underside for a bottom bee space hive need to be tailored to the hive wall thickness in a similar manner to crown boards, queen excluders, etc. Our own choice is the Ashforth type because if the hive has a slight tilt the bees can consume the whole lot without leaving a small pool of syrup in the feeder.

There is much merit in adopting the method and equipment used by Bro. Adam whereby he had a combined crown board/feeder on each of his colonies, thereby saving one piece of equipment per hive and always having a feeder in place for use in an emergency (which I don't think he ever had!).

Butler cage

A careful perusal of Butler cages for sale from some of the equipment suppliers reveals that the wire mesh is too fine and the cross-sectional dimensions are incorrect.

The mesh needs to be about 12 per inch in approximately 25 swg (standard wire gauge) in order to allow the worker bees to groom and feed the queen through the mesh; a necessary part of the queen introduction process.

The cage should have cross-sectional dimensions of ½ inch x ¾ inch and have a length of about 3 to 4 inches. It is the ½ inch dimension that is critical as it represents two bee spaces and therefore just fits snugly between two frames in the brood chamber. All our own Butler cages have had a 1½ inch long panel pin soldered into the cage at the closed end with 1 inch protruding from the cage; this is used for pushing into the comb when introducing a queen. It is better and easier than pressing the cage into the comb.

Extractors

It is not proposed to discuss extractors in detail but just to sound a few words of warning when buying this very expensive bit of equipment. It is a bit of puzzle why radial extractors are available in sizes to take 9, 10, 12, 15, 20, or 24 super frames let alone the smaller tangential types with and without motors, with or without reversing gear. The speeds differ from maker to maker from approximately 280 rpm to 400 rpm. Surely there must be an optimum specification to the advantage of most beekeepers?

Most hobbyist beekeepers using Manley frames with 10 per super would be seeking a 10 frame extractor. That was my requirement years ago and having purchased the machine it was found that it would not accommodate 10 Manley frames but would accommodate 10 super frames with narrow side bars. The result is that all my Manley frames have been modified to fit 10 into the extractor, all 600 or so of them! Caveat emptor.

This is another aspect of beekeeping equipment that requires examination and a specification prepared for manufacturing purposes.

Brood boxes

With the build up of propolis at the ends of the top bars in particular, and at other places in the brood box it becomes necessary to take the brood boxes out of service at regular intervals for scraping, cleaning, repair where necessary, flaming and creosoting. In general, it is necessary to do this once every two years; the best time to do this is in the spring before there are any supers on the colony, ie. at the first inspection after floor boards have been changed when the queen is being marked and the colony has its first seasonal assessment. Continual scraping with a sharp triangular scraper soon removes enough wood (especially if it is a soft wood like cedar) from the inside of the hive wall, adjacent to the ends of the frame top bars, to make the end play unacceptably large and start creating an unacceptable build up of propolis. The distance between the inside walls on a Modified National should be 17$\frac{1}{16}$ inches, ie. $\frac{1}{16}$ inch greater than the length of the top bars. These measurements vary depending on the hive type.

A simple method of getting over this problem is to scrape one side clean back to the wood and glue in a strip of Formica to take up the slack ensuring that the inside dimension is still 17$\frac{1}{16}$ inches. A few years later when the other side gets scraped away then a second piece of Formica can be fitted. The Formica of course gets its fair share of propolis but it scrapes off very easily without wear to its surface. Furthermore, the Formica can withstand the heat of scorching when the box is flamed.

This simple remedy prompts the question whether some device such as described should be incorporated into a hive specification and fitted when the hive is newly constructed.

Beeswax foundation

Beeswax foundation can be considered as part and parcel of beekeeping equipment; thousands of sheets are purchased every year both wired and unwired all eventually used, presumably successfully, in the colonies of the beekeepers concerned. There is no known specification for this foundation, so one may reasonably ask why bother about it?

There are basically two types of foundation, namely, thick for use in the brood chamber and thick enough to embed a wire into it without the wire being exposed on either side and the other thin for use in supers and sections. The latter is a food, and is eaten in section form or if the super is used for cut comb.

The question now arises how thick should the thick be and how thin should the thin be and how should this be measured and marketed? Traditionally, thick foundation has been measured by weight and by the number of sheets, 8 sheets per pound weight for British National frames and 5 sheets per pound for Modified Commercial frames. The thin is the real problem and here I shall relate a short story to illustrate the problem. Some years ago we were buying foundation for supers (we require it as thin as possible to avoid, what is known as, the fish bone effect in the middle of cut comb if it is too thick) and 300 sheets were purchased from Exeter Bee Supplies. Supers were fitted with the foundation and the bees refused to work them. Some test supers were fitted, half with the new foundation and half with some older foundation (but thicker). The bees worked the 5 frames of the older foundation but refused to touch the 5 frames of new. Further careful inspection of the 'duff' foundation revealed that it was very pliable when cold and did not have that characteristic aroma of new wax foundation; in fact it was almost odourless. After discussion with Exeter Bee Supplies they agreed that there was something wrong with the foundation and disclosed that it had been purchased by them from Thorne. EBS, of course, replaced the foundation with some other they had in stock. I was convinced that the foundation had in some way become adulterated with something; but what? I could not find out and the rest is surmise.

A couple of years later my daughter and her husband lost a colony of worms (the eco-friendly system of disposing of household vegetable waste in a 'wormery') after putting in the usual amount of newspaper. However, in this case it was not normal newspaper but the yellow sheets from an old 'Yellow Pages'. The pages were toxic to the worms and unfortunately the worms could not avoid them. I have been informed that the yellow dye used to colour the yellow pages could be the toxin that killed the worms.

Thorne sells coloured foundation for making candles, the colours range from black, red, green and various shades of yellow. Could some packets of foundation which had been yellow dyed have been inadvertently mixed up with packets of foundation for supers? Discussion by telephone with Thorne was inconclusive, such a mix up could never happen, so we will never know the truth of the matter but it illustrates two things; the first is the bees were clever enough to reject it out of hand and the second was it was something destined for the food chain.

My own opinion in the matter is that there should be a specification for beeswax foundation but it will either be a very tricky one to write or a very simple one and I visualise a lot of opposition from the few foundation makers in the UK.

Porter bee escape

The Porter bee escape has been around now for a very long time and is a prime example of equipment that is much worse than its forbears. I have two made from tin plate, each being a four way device, yet still the traditional size and shape. It is divided longitudinally and has four phosphor bronze springs soldered to the inside walls with ⅛ inch gap between the end of the spring and the dividing wall. They have been used since after the last war and never fail. The secret is in the phosphor bronze spring being just the right dimensions and set with just the right amount of tension for the bee to open easily without being able to return.

Most Porter bee escapes are now made of plastic with very inadequate springs; in general, the springs are too large and too heavy to do their job properly and their fixing inside the body of the escape is crude beyond words. Some years ago Steele and Brodie sold a plastic Porter bee escape with removable plastic springs which was well engineered but, alas, the springs let it down and it was not reliable in use.

A little thought and it will readily be agreed that the spring arrangement needs specifying for the gadget to work satisfactorily.

Other items

There are many other items of beekeeping equipment that are either unsatisfactory or which could be improved and are not directly related to either bee space or the minimum propolis criterion. Two have been briefly discussed above as examples. Other items include, standard nucleus hives, mini nucs, queen rearing equipment, pollen traps, mesh floors for Varroasis and wintering, solar wax extractors, etc., etc.; the list is endless.

PROPOSALS

It will be clear from the forgoing that little progress has been made in rigorously designing beekeeping equipment to comply with the design criteria enunciated earlier. The reason must be that no one, curiously enough, has taken the trouble to postulate any design criteria. However, it has been shown that in the short term a simple bit of DIY can make enormous differences to minimising the problems caused by the building of brace/burr comb and propolising joints throughout the hive. It will also be clear that other beekeeping equipment has some serious shortcomings. With a little thought and a bit of brain-storming it should be possible to make some progress to improve the situation for future generations of beekeepers. The next item to be addressed is how to proceed to effect some improvements?

Standards and the organisation

Standards or specifications? A standard is defined as 'a definite level of excellence or adequacy required' whereas a specification is defined as 'a detailed description of requirements'. In the past, that is, up to 1984 standards were used and in particular those of the British Standards Institute (BSI) with the exception of the BBKA standard for a nucleus and a stock of bees. As we are dealing, in the main, with equipment it is suggested that 'specifications' should be adopted rather than standards but it is of little consequence providing something is written down for all to work to and so that everyone is facing the same way with a common objective.

There is no doubt that the only organisation which should be responsible for any specifications to do with beekeeping equipment in this country should be the BBKA, a national organisation. There are other organisations such as the Bee Farmers Association (BFA) who are major users of equipment and, of course, the equipment manufacturers themselves but neither of these, in my opinion, are likely to have the motivation to do anything about the problem. There are reasons for this of course; being a bee farmer requires an inordinate amount of effort to earn a living keeping bees and few, if any, would wish to devote further time to preparing specifications for the future. I hope that I am not maligning the bee farmers in this matter. The equipment manufacturers will not wish to rock the boat and my guess is that they will wish to maintain the status quo and will be reluctant to make any changes unless they are called for by a national body. It will be argued that Scotland, Wales and Ireland should be part of the act. In the ultimate I agree, in the short term I disagree; it would be better to get things launched and set up before widening the scope and it would be very easy for them to adopt any standard or specification that took their fancy.

We must look seriously at the BBKA; the graph below shows the membership figures from the early 80s to the present time. My timely warning in 2002 that it was going out of business in 2017 (by extrapolation from 1982 to 2001) appears to have been heeded; or was this just coincidence? One should remember that just after the war in the late 40s the membership was of the order of 80,000. Also, we must remember that the BBKA abandoned all standards in 1984. Are the chances good of the revival continuing after the massive downward trend over 50 years or more? I think that there is a fair possibility and on this basis equipment standards/specifications are essential for the future if good husbandry is to be attained.

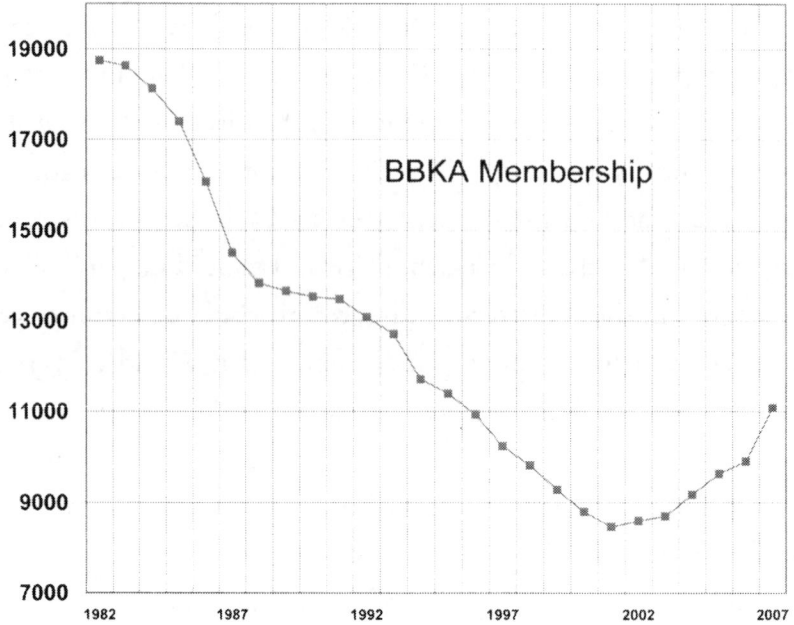

There is one other possibility and that is DEFRA; their forebears, MAFF, were responsible for the design of the Modified National hive just after the last war. Judging them on their past performance namely, CAP, salmonella, BSE, foot & mouth in the agricultural sector and Varroasis in the beekeeping sector they are, in my opinion, unlikely to be short listed as a possible starter.

It looks as though it must be the BBKA unless other ideas are put forward.

Committees and those on them

At the time of writing and taken directly from BBKA News No. 134, there are 5 Committees as follows:

Education and Husbandry,
Examination Board,
Finance,
Publicity and Promotions,
Technical

together with one working party on Constitution Revue (a loosely constructed theatrical show!); I think they mean review!

Should any of the above Committees be responsible for producing specification/standards for approval and issue by the Executive of the BBKA? In my opinion there are only two which could be considered and that is the Education and Husbandry or the Technical depending on their present terms of reference to which I am not party. Otherwise another working party responsible to whom? Directly to the Executive or to one of the two Committees?

It is my considered opinion that another working party is the answer, the smaller the better. The constitution should be a representative of the equipment manufacturers freely put forward by their own group and not part of the BBKA hierarchy, one of the BBKA Executive responsible for the working party budget and one other BBKA member with professional engineering qualifications together with extensive beekeeping experience. No, I don't want the job but they may use this monograph as much as they wish and copy it to their hearts content, with prior permission, if that is helpful.

CONCLUSIONS

In the long term, the solution must rest with the BBKA to resurrect the standards and specifications for frames and other equipment and then encourage the suppliers to manufacture accordingly. The ones who suffer most at the present time are those new to beekeeping, they inherit the problems of the past which should have been discarded many years ago.

I hope that the foregoing dissertation will prove to be helpful in the short term for beekeepers to modify their own equipment to make the manipulation of their hives easier, and therefore, more enjoyable. In the long term, I hope the BBKA will take up the challenge of preparing well written specifications and standards for all to use.

If all else fails, then I hope that the membership will take up the matter through their branches, association meetings and finally passing motions at the ADM enforcing the Executive to take action.

The present situation, in my opinion, remains a disgrace.

IMPERIAL/METRIC CONVERSIONS

Imperial (fraction) -inch	Imperial (decimal) - inch	Metric - mm
1/32	0.031	0.8
1/16	0.062	1.6
1/8	0.125	3.2
1/4	0.250	6.4
5/16	0.312	7.9
3/8	0.375	9.5
7/16	0.438	11.1
1/2	0.500	12.7
9/16	0.563	14.3
11/16	0.688	17.5
3/4	0.750	19.1
1	1.000	25.4
1 1/16	1.062	27.0
1 3/8	1.375	34.9
1 9/20	1.450	36.8
1 1/2	1.500	38.1
1 5/8	1.625	41.3
1 7/8	1.875	47.6
2	2.000	50.8

Lightning Source UK Ltd.
Milton Keynes UK
UKOW010241170113

204957UK00004B/131/P